COMPLETE GUIDE TO PANCREATIC CANCER

Comprehensive Resource To Diagnosis, Treatment Options, Survival Strategies, Nutritional Support For Patients And Families

DEHART HAIRSTON

© [DEHART HAIRSTON], [2024]

All rights reserved. No part of this publication may be reproduced, distributed, or transmitted in any form or by any means, including photocopying, recording, or other electronic or mechanical methods, without the prior written permission of the publisher, except in the case of brief quotations embodied in critical reviews and certain other noncommercial uses permitted by copyright law.

DISCLAIMER

This book's content is only intended for general informative purposes. At the time of writing, the author has taken every precaution to guarantee that the material is correct and current. Nevertheless, the author disclaims all explicit and implicit representations and guarantees about the availability, appropriateness, correctness, completeness, and usefulness of the material on these pages.

Since the author is not a licensed medical practitioner, the material in this book shouldn't be interpreted as medical advice. Before making any modifications to their diet, exercise regimen, or medical treatment, readers are urged to speak with a licensed healthcare provider.

Moreover, the author has no connection to any of the businesses, organizations, or people that are discussed in this book. Any mentions of goods, services, businesses, or people are purely informative and do not indicate endorsement or suggestion.

This book's content is entirely dependent on the author's expertise, study, and comprehension of the topic. Despite having taken reasonable care to offer correct information, the author disclaims all liability for any mistakes or omissions in the material as well as for any losses, harm, or damages resulting from using the information.

It is recommended that readers use their own judgment and discretion when applying the knowledge in this book to their own situations. The use or implementation of any material in this book may result in unfavorable repercussions, directly or indirectly, for which the author assumes no liability.

By reading this book, you agree to release and hold the author harmless from any claims, losses, liabilities, costs, or expenditures resulting from or related to the use of the information you get from it.

Table of Contents

CHAPTER 1 13
Understanding Pancreatic Cancer 13
What Is Pancreatic Cancer? 13
Types And Stages Of Pancreatic Cancer 14
Risk Factors And Causes.......................................15
1. Age:..15
2. Tobacco Use:..15
3. Family History: 15
4. Chronic pancreatitis: 16
5. Obesity and Diabetes: 16
6. Dietary Factors:...16
7. Exposure to Specific Chemicals:........................16
8. Race & Ethnicity:...17

CHAPTER 2 19
Signs And Symptoms 19
Recognizing Early Signs 19
Common Symptoms And Their Impact 20
When To Seek Medical Help 22

CHAPTER 3 25
Diagnosis And Screening 25

- Diagnostic Tests And Procedures 25
- Importance Of Early Detection 27
- Screening Guidelines 28
- **CHAPTER 4** 31
  - Treatment Options 31
  - Surgery, Chemotherapy, And Radiation Therapy 31
    - Surgery: 31
    - Chemotherapy: 32
    - Radiation Treatment: 32
  - Targeted Therapy And Immunotherapy 33
    - Personalized Medicine: 33
    - Immunotherapy: 34
  - Palliative Care And Supportive Therapies 35
    - Hospice Care: 35
    - Complementary Medicines: 36
- **CHAPTER 5** 39
  - Nutritional Support 39
  - Importance Of Nutrition In Pancreatic Cancer 39
  - Diet Recommendations And Guidelines 40
    - • Consumption of Protein: 40
    - • Caloric Needs: 40

- Hydration: 41
- Vitamins and Minerals: 41

Managing Symptoms Through Diet 42
- Nausea and Vomiting: 42
- Loss of desire: 42
- Digestive Problems: 43
- Tiredness: 43

CHAPTER 6 45
Coping Strategies 45
Emotional Impact Of Pancreatic Cancer 45
Support Systems And Resources 46
Coping Mechanisms For Patients And Caregivers 48

CHAPTER 7 51
Clinical Trials And Research 51
Understanding Clinical Trials 51
Current Research And Breakthroughs 54
How To Participate And Access Experimental Treatments 56

CHAPTER 8 61
Survivorship And Quality Of Life 61
Life After Treatment 61

Managing Long-Term Side Effects 62
Maintaining Quality Of Life And Well-Being 63
CHAPTER 9 65
Caregiver's Guide 65
Roles And Responsibilities Of Caregivers 65
Self-Care Tips For Caregivers 67
Support Networks For Caregivers 70
CHAPTER 10 73
Advocacy And Awareness 73
Spreading Awareness About Pancreatic Cancer ... 73
Advocating For Research Funding 75
Getting Involved In Pancreatic Cancer Organizations 76
CONCLUSION 79
THE END 82

ABOUT THE BOOK

For anybody affected by this illness, "Pancreatic Cancer" is more than simply a book—it's a lifeline. Within its pages is a thorough handbook that clarifies every facet of pancreatic cancer, from comprehending its intricacies to navigating available treatments, and providing support to both patients and caregivers.

The foundation is laid in Chapter 1, which offers a thorough explanation of the kinds, stages, risk factors, and causes of pancreatic cancer. For early identification and well-informed decision-making, this information is essential.

An in-depth discussion of the often subtle indications and symptoms is covered in Chapter 2, enabling readers to identify them early and take proactive measures toward diagnosis and treatment. It can save lives to know when to seek medical attention.

In Chapter 3, diagnosis and screening are covered. The significance of early detection using a variety of tests and treatments is emphasized. By following the screening standards, people may take charge of their health and identify their condition early on.

With its exploration of therapeutic options ranging from surgery to targeted therapy and palliative care, Chapter 4 provides a ray of hope. Equipped with this understanding, patients may have knowledgeable conversations with their healthcare professionals and choose the most appropriate course of action.

To manage pancreatic cancer, Chapter 5 emphasizes the vital role that nutrition plays. It provides helpful dietary suggestions and techniques to reduce symptoms and enhance general health.

In Chapter 6, coping mechanisms for patients and caregivers are discussed, along with the emotional toll that the condition takes. Throughout the trip, this area is essential for preserving both mental and emotional resilience.

In Chapter 7, readers are given an overview of clinical trials and research, along with chances to participate in experimental therapies and learn about recent advances in the field.

After the dust of therapy settles, Chapter 8 focuses on quality of life and survivability, looking beyond treatment. It gives patients the skills they need to handle long-term adverse effects and navigate life after treatment.

In Chapter 9, the critical role of caregivers is acknowledged, and advice and assistance are provided to those who give selflessly to take care of their loved ones. Caregivers may prioritize their well-being while assisting their loved ones by using helpful advice and tools.

In Chapter 10, the value of awareness-raising and activism is emphasized, encouraging readers to take up arms against pancreatic cancer. The landscape of pancreatic cancer treatment and research may be significantly changed by people via raising awareness and supporting financing for research.

"Pancreatic Cancer" is more than just a book; it's a road map that provides direction, encouragement, and hope at every turn on an arduous trip. This is an essential book for anybody dealing with pancreatic cancer, whether they are a patient, caregiver, or advocate.

CHAPTER 1

Understanding Pancreatic Cancer

Pancreatic cancer is a very dangerous disease that is difficult to cure since it is often discovered after it has progressed. Patients, caregivers, and healthcare professionals need to understand this illness. Let's examine the definition, kinds, stages, and underlying causes of pancreatic cancer in detail.

What Is Pancreatic Cancer?

The cause of pancreatic cancer is the uncontrolled growth of aberrant cells in the pancreas, an important organ situated below the stomach. By generating enzymes and insulin, the pancreas plays a critical role in digestion and blood sugar management. The pancreas' regular function is disrupted and there is a risk of the malignant cells spreading to other regions of the body when they multiply there.

Types And Stages Of Pancreatic Cancer

Pancreatic cancer comes in several forms, each with unique traits and therapeutic modalities. Adenocarcinoma is the most prevalent form, making up around 90% of instances of pancreatic cancer. Usually, adenocarcinomas start in the cells that line the pancreatic ducts.

Pancreatic cystic tumors, which originate as fluid-filled sacs in the pancreas, and neuroendocrine tumors (NETs), which grow from hormone-producing cells in the pancreas, are less frequent forms. Determining the best course of therapy requires an understanding of the particular form of pancreatic cancer.

According to the degree of its dissemination, pancreatic cancer is also staged. Healthcare professionals may evaluate the disease's severity and adjust their treatment plans with the aid of the staging system. The stages are 0 through IV, where stage 0 denotes cancer that has only progressed to the pancreatic ducts' innermost layer and stage IV denotes cancer that has migrated to other organs.

Risk Factors And Causes

Pancreatic cancer risk may be elevated by several variables. The following are a few of the most important risk factors:

1. **Age:** Most instances of pancreatic cancer are discovered after the age of 65, and the disease is more frequent in older persons.

2. **Tobacco Use:** One of the main risk factors for pancreatic cancer is cigarette smoking. Compared to non-smokers, smokers have a minimum of a twofold increased risk of developing the illness.

3. **Family History:** People who have a genetic condition like Lynch syndrome or hereditary

pancreatitis, or who have a family history of pancreatic cancer, are more likely to have the illness.

4. **Chronic pancreatitis:** Prolonged pancreatic inflammation often brought on by alcoholism or other variables, might eventually increase the risk of pancreatic cancer.

5. **Obesity and Diabetes:** Type 2 diabetes and being overweight or obese are linked to an increased risk of pancreatic cancer.

6. **Dietary Factors:** Eating a diet heavy in processed and red meats and low in fruits and vegetables may increase your chance of developing pancreatic cancer.

7. **Exposure to Specific Chemicals:** Certain chemicals, including petrochemicals, dyes, and

pesticides, may increase an individual's chance of developing pancreatic cancer.

8. Race & Ethnicity: Compared to those in other racial and ethnic groups, African Americans have a higher risk of developing pancreatic cancer.

People may lower their chance of having pancreatic cancer by being proactive and learning about these risk factors and causes. This might include changing one's way of living to include things like giving up smoking, keeping a healthy weight, and eating a well-balanced diet full of fruits and vegetables. Regular medical examinations and screenings may also aid in the early detection of pancreatic cancer when therapy is most successful.

CHAPTER 2

Signs And Symptoms

Recognizing Early Signs

Effective therapy and better results from pancreatic cancer treatment depend on early identification. It might be difficult to recognize the symptoms, however, since they often resemble those of other, less serious illnesses. However, several early warning signs need to spur further research.

Jaundice, or yellowing of the skin and eyes, is one of the first indications of pancreatic cancer. This happens when the tumor blocks the bile duct, which causes bilirubin to accumulate in the blood. Moreover, inexplicable weight loss may indicate pancreatic cancer, especially if it is accompanied by weariness and appetite loss.

This often happens as a consequence of the cancer obstructing the body's normal food digestion process.

Additional early warning indicators might include varying degrees of discomfort or pain in the abdomen that could spread to the back. Due to inadequate pancreatic enzyme production, some people may also notice changes in their bowel habits, such as diarrhea or pale, oily stools.

It's crucial to remember that these symptoms may also be brought on by several other illnesses and are not specific to pancreatic cancer. Nonetheless, it is essential to speak with a healthcare provider for a more thorough assessment if they continue or become worse over time.

Common Symptoms And Their Impact

Additional symptoms may appear as pancreatic cancer advances, significantly impairing a person's health and quality of life. Depending on the tumor's size, location, and ability to spread to other organs, these symptoms may differ.

One typical sign of pancreatic cancer is persistent stomach pain, which is commonly characterized as a dull aching or discomfort that becomes worse after eating or lying down. This discomfort may be accompanied by nausea, vomiting, and trouble breaking down meals, which might result in malnourishment and debility.

When a tumor blocks the pancreatic duct, people may get pancreatitis, an inflammation of the pancreas that may lead to high fever, stomach issues, and agonizing pain in the abdomen. Furthermore, when the cancer spreads and develops, it may put strain on surrounding tissues and organs, causing symptoms including jaundice, back discomfort, and trouble breathing.

These symptoms may have a significant negative influence on many aspects of everyday living, including mobility, emotional stability, food, and sleep. Because of this, people exhibiting these symptoms must get medical help as soon as possible to have a correct diagnosis and suitable therapy.

When To Seek Medical Help

Managing possible pancreatic cancer symptoms requires knowing when to get medical attention. Even while many of the disease's symptoms may also be seen in less severe diseases, several warning signs should be taken seriously and should need quick attention.

Seeking quick medical attention is crucial if you have chronic jaundice, which is characterized by yellowing of the skin and eyes. Jaundice necessitates further testing since it may be a sign of pancreatic cancer or some underlying problem with the liver or bile duct.

It's important to pay attention to unexplained weight loss, especially if it's accompanied by additional symptoms like nausea, stomach discomfort, or changes in bowel habits. Unintentional or rapid weight loss should be evaluated medically since it may indicate cancer or other illnesses.

Similarly, you should see your doctor if you have ongoing stomach discomfort, particularly if it is severe or becoming worse over time. Even though there are many possible reasons for stomach discomfort, such as gastrointestinal disorders and musculoskeletal conditions, pancreatic cancer must be ruled out, especially if additional symptoms are present.

In general, you should get medical attention right away if you have any worrisome symptoms that last longer than a few weeks or have a substantial effect on your day-to-day activities. Individuals with pancreatic cancer must have an early diagnosis and treatment, so don't be afraid to voice your concerns to a medical practitioner.

CHAPTER 3

Diagnosis And Screening

Diagnostic Tests And Procedures

Several tests and techniques are often used to diagnose pancreatic cancer to find anomalies in the pancreas or its surrounding tissues. Imaging examinations like computed tomography (CT), magnetic resonance imaging (MRI), or ultrasound are often one of the first stages. By using these imaging methods, medical professionals may see the pancreas and surrounding tissues and search for any anomalies or malignancies.

Blood tests may be performed to measure levels of certain compounds, such as tumor markers like CA 19-9, that may suggest pancreatic cancer in addition to imaging studies. It's crucial to remember that increased levels of these markers may also be linked to other illnesses, so they should only be seen as one piece of the diagnostic jigsaw rather than as conclusive evidence of cancer.

A biopsy might be done for a more thorough analysis. A pathologist will analyze a tiny sample of tissue removed from the pancreas or a suspected tumor under a microscope during a biopsy. This validates the existence of malignant cells, enabling a conclusive diagnosis of pancreatic cancer.

Pancreatic cancer diagnosis may also be made via endoscopic techniques. Using endoscopy and ultrasound, endoscopic ultrasonography (EUS) creates finely detailed pictures of the pancreas and its surrounding tissues. Fine-needle aspiration (FNA), a technique that involves inserting a tiny needle into the pancreas to remove cells for analysis, is another way that tissue samples may be obtained using this operation.

Importance Of Early Detection

For pancreatic cancer to improve prognoses and expand treatment choices, early identification is essential. Early-stage pancreatic cancer often shows minimal symptoms, and by the time symptoms show up, the disease may have migrated to other bodily areas, making treatment more challenging.

Pancreatic cancer may present with nonspecific symptoms such as stomach discomfort, weight loss, jaundice, and digestive problems, making early diagnosis difficult. Diagnoses might be delayed since several other illnesses could be causing these symptoms.

Nonetheless, improvements in early detection rates may be attributed to advancements in diagnostic procedures and heightened knowledge of risk factors. A more manageable stage of the illness may be identified by screening high-risk people, such as those with a family history of pancreatic cancer or certain genetic abnormalities.

Screening Guidelines

Because pancreatic cancer is not as common as other cancers and there are no reliable screening tools, the process of screening for the disease is more complicated. There aren't any generally acknowledged screening protocols for the general public at this time.

On the other hand, as part of a surveillance program, those with recognized risk factors for pancreatic cancer could have screening tests. This usually involves blood tests to check for changes in tumor markers over time, as well as imaging studies like CT or MRI scans.

The criteria for identifying high-risk people might change based on several variables, including genetic predisposition, age, and family history.

People who have risk factors should talk to their healthcare professional about screening choices to decide what is best for them.

All things considered, while screening for pancreatic cancer is still difficult, technological and scientific advancements are making it easier to identify the illness early on, which benefits patients. Finding pancreatic cancer in its earliest and most curable stages requires proactive conversations with healthcare professionals, regular check-ups, and knowledge of risk factors.

CHAPTER 4

Treatment Options

Surgery, Chemotherapy, And Radiation Therapy

Surgery:

For pancreatic cancer, surgery is often the first line of therapy, particularly if the tumor is confined and hasn't progressed to other organs. The purpose of surgery is to remove the tumor together with any surrounding tissues that are impacted. A Whipple surgery (pancreaticoduodenectomy) is a standard surgical treatment for pancreatic cancer that involves removing the gallbladder, bile duct, and a portion of the small intestine. The goal of this lengthy procedure is to remove the tumor entirely and guarantee that no malignant cells remain.

Chemotherapy:

Strong medications are used in chemotherapy to either kill or inhibit the growth of cancer cells. Neoadjuvant chemotherapy is often administered before surgery to reduce the tumor's size and facilitate its surgical removal. Chemotherapy may be administered after surgery to treat cancer that has spread outside of the pancreas or to eradicate any cancer cells that may still be present. Chemotherapy is generally given in cycles to give the body time to recuperate between treatments. It may be delivered intravenously or orally.

Radiation Treatment:

High-energy radiation is used in radiation treatment to target and kill cancer cells. When treating pancreatic cancer, it's often used in conjunction with surgery and/or chemotherapy. The most popular kind of radiation treatment for pancreatic cancer is called external beam radiation therapy, in which radiation is directed toward the tumor by a machine that is outside the body. Radiation therapy may be used as a palliative measure to reduce symptoms and enhance quality of life, or it can be used before surgery to reduce the tumor and after surgery to eradicate any cancer cells that may still be present.

Targeted Therapy And Immunotherapy

Personalized Medicine:

One kind of treatment called targeted therapy focuses on certain chemicals that are essential to the development and metastasis of cancer cells. Targeted treatment is intended to selectively target cancer cells while limiting harm to healthy cells, in contrast to chemotherapy, which damages all rapidly dividing cells. EGFR and VEGF are two examples of proteins or pathways that targeted therapy medications may target for pancreatic cancer. These treatments aim to block the overactivity of these proteins or pathways in cancer cells. Chemotherapy is one treatment that targeted therapy may be administered with or without.

Immunotherapy:

One kind of cancer treatment that boosts the body's immune system to combat the disease is immunotherapy. It works by strengthening the body's defenses against disease or by teaching the immune system to identify and target cancerous cells. Immunotherapy is still being researched for its potential to treat pancreatic cancer, despite its success in treating other forms of cancer. Clinical studies are being conducted on some immunotherapy medications, such as immune checkpoint inhibitors, to treat pancreatic cancer. These medications function by unblocking the immune system, enabling it to more effectively identify and combat cancerous cells.

Palliative Care And Supportive Therapies

Hospice Care:

The goal of palliative care is to alleviate the stress and discomfort associated with a life-threatening condition, such as pancreatic cancer. It may be given in conjunction with curative therapies to enhance quality of life, not simply for end-of-life care. Palliative care may include symptom control, pain treatment, emotional support, and help with day-to-day tasks including paperwork and finances. Throughout the cancer journey, palliative care teams collaborate closely with patients, their families, and other medical professionals to meet physical, emotional, and spiritual needs.

Complementary Medicines:

Supportive therapies may be very important in controlling pancreatic cancer and its impact on general health in addition to medical treatments. Physical therapy to increase mobility and function, psychological assistance to deal with the emotional repercussions of cancer diagnosis and treatment, and dietary counseling to help preserve strength and manage the side effects of treatment are some examples of these treatments. By addressing the needs of the whole person rather than simply the illness, supportive therapies seek to improve the quality of life and provide comprehensive treatment.

Patients and their loved ones may make educated choices about their care and collaborate with healthcare professionals to create individualized treatment plans that best suit their needs and objectives by being aware of the many treatment options available for pancreatic cancer.

Every treatment method has advantages and possible drawbacks, and the decision on which to get depends on several variables, including the cancer's stage, general health, and personal preferences. To feel secure in their treatment choices, patients must have a full discussion of their alternatives with their medical team and express any questions they may have.

CHAPTER 5

Nutritional Support

Importance Of Nutrition In Pancreatic Cancer

Pancreatic cancer treatment heavily depends on nutrition. Patients often have a variety of symptoms that might interfere with their ability to consume and absorb nutrition. Sustaining an optimal diet is crucial for bolstering the immune system, encouraging recovery, and controlling the negative effects of medical interventions.

Fatigue, malnourishment, and weight loss are just a few of the side effects that may result from pancreatic cancer and its therapies. In addition to raising the general quality of life and maybe increasing treatment results, proper eating can help lessen these consequences. Sufficient nourishment facilitates the body's capacity to endure medical interventions, recuperate after surgical procedures, and manage the rigors of cancer treatment.

Diet Recommendations And Guidelines

The goal of food recommendations for people with pancreatic cancer is to maintain sufficient nutrition levels while controlling symptoms and adverse effects. To ensure that the body gets all the nutrients it needs throughout treatment, a balanced diet full of different nutrient-rich foods is crucial.

- Consumption of Protein: Protein is essential for immune system maintenance and tissue repair. Increasing protein consumption may assist people with pancreatic cancer in maintaining their muscle mass and avoiding malnourishment. It is advised to consume lean protein sources such as fish, chicken, tofu, and lentils.

- Caloric Needs: Getting enough calories may be difficult for people with pancreatic cancer, particularly if they have trouble eating or lose their appetite. Large meal portions are not necessary when consuming calorie-dense meals and snacks to assist enhance energy consumption. Nuts, avocados, and olive oil are examples of healthy fats that may provide important calories.

- **Hydration:** Patients with pancreatic cancer often have symptoms including nausea, vomiting, and diarrhea. These symptoms may be managed by maintaining enough hydration. Maintaining an appropriate fluid balance may be facilitated by drinking water throughout the day and choosing hydrating meals like fruits and vegetables.

- **Vitamins and Minerals:** Several vitamins and minerals are essential for maintaining general health and immunological function. Supplementing these nutrients or including foods high in them in their diet may be beneficial for people with pancreatic cancer.

Dairy products, fruits, vegetables, and whole grains are great providers of important vitamins and minerals.

Managing Symptoms Through Diet

Dietary changes may be used to control or reduce a number of symptoms related to pancreatic cancer and its therapies.

- **Nausea and Vomiting:** You may lessen nausea and vomiting by eating small, frequent meals and steering clear of foods that are hot, fatty, or too sweet. Other potential remedies include ginger, peppermint, and simple carbs like bread or crackers.

- **Loss of desire:** Eating may be stimulated by providing palatable, nutrient-rich meals to pique desire. Meals may be made more delightful by including favorite foods, experimenting with various textures and tastes, and dining in a comfortable setting.

- **Digestive Problems:** Patients with pancreatic cancer may have constipation or diarrhea as digestive problems. It may be beneficial to consume soluble fiber from foods like bananas, apples, and oats to help control bowel motions. Steer clear of items that cause gas, such as cabbage and beans.

- **Tiredness:** Consuming well-balanced meals and drinking enough of water will help prevent pancreatic cancer-related tiredness. While consuming modest quantities of caffeine from tea or coffee could provide you with a short-term energy boost, it's important to limit your caffeine consumption.

Through adherence to these dietary guidelines and suggestions, patients with pancreatic cancer may successfully control their symptoms, meet their nutritional requirements, and enhance their general health and well-being throughout treatment. It is possible to guarantee that nutritional therapies are customized to each patient's requirements and preferences by collaborating closely with a healthcare team that includes a certified dietitian.

CHAPTER 6

Coping Strategies

Emotional Impact Of Pancreatic Cancer

For patients and their loved ones, receiving a pancreatic cancer diagnosis may be an extremely taxing and upsetting process. A variety of feelings, such as dread, grief, rage, and worry about the future, might be triggered by the news. It's common to feel overburdened and to grieve for the life you led before receiving the diagnosis.

Pancreatic cancer's image as a very aggressive and sometimes late-stage illness is one of the main emotional obstacles associated with the disease. Feelings of helplessness or despair may result from this. But it's important to keep in mind that each person's experience with pancreatic cancer is different, and there's always hope for a better prognosis and more successful therapy.

It's critical to have emotional support throughout this period. Speaking with loved ones, friends, or a mental health professional may provide a safe space to vent emotions and find solace. Joining a support group for individuals with pancreatic cancer and their caregivers may also provide invaluable companionship and understanding from those going through similar circumstances.

Support Systems And Resources

Having a strong support network and access to a variety of services are typically necessary for navigating the pancreatic cancer journey. Numerous people may provide support, including friends, family, medical experts, and neighborhood groups. In addition to providing patients with emotional support and practical help with everyday duties, loved ones may also go with patients to appointments.

Social workers, nurses, and oncologists are among the healthcare professionals who are vital in helping patients find services and provide assistance. They may provide advice on palliative care services, symptom management, and possible courses of therapy. To lessen the financial burden of cancer treatment, social workers may also help with accessing financial assistance programs, transportation services, and other useful resources.

Patients and caregivers with pancreatic cancer have access to a plethora of Internet and community-based services in addition to their support networks. These might be things like learning resources, social networking sites, and groups that help people find their way around the medical system and find available treatments.

Coping Mechanisms For Patients And Caregivers

Adopting useful coping techniques and being resilient are necessary for managing pancreatic cancer. Throughout therapy, it may be beneficial for patients and caregivers to investigate different approaches to stress management and mental well-being.

Both patients and caregivers must practice self-care. This might be taking part in joyful and calming pursuits like meditation, physical activity, hobbies, or time spent in nature. Burnout may be avoided and general wellbeing can be enhanced by setting aside time for personal needs and taking breaks from caring obligations.

Coping with pancreatic cancer requires open communication. Open and sincere communication is essential between patients, caregivers, and healthcare professionals. Relationships may be strengthened and a feeling of mutual support can be fostered by sharing worries, anxieties, and objectives.

It's crucial to keep your expectations reasonable and your attention on the here and now. Although worrying about the future is normal, maintaining awareness of the here and now may help ease tension and encourage serenity. Establishing modest, attainable objectives might give one a feeling of advancement and success.

Finally, it's important to seek expert assistance when required. Therapists and counselors who specialize in mental health issues may provide coping mechanisms for managing stress, anxiety, and sadness. Support groups may provide a feeling of belonging and affirmation of shared experiences for both patients and caregivers.

Through the use of these coping methods and the utilization of existing support networks and resources, individuals with pancreatic cancer may effectively confront its difficulties with fortitude and optimism.

CHAPTER 7

Clinical Trials And Research

Understanding Clinical Trials

In the medical profession, clinical trials are essential, particularly when fighting illnesses like pancreatic cancer. By evaluating the security and effectiveness of novel medications and treatments, these studies act as a link between scientific advancement and practical implementation. Clinical trials are essentially scientific investigations conducted on human subjects to evaluate the efficacy of novel medications, therapies, or methods.

The main objective of pancreatic cancer clinical trials is to enhance patient outcomes via the development of more effective disease prevention, detection, and treatment strategies.

These studies might test novel chemotherapeutic medications, investigate targeted treatments, or assess cutting-edge surgical methods, among other topics. Clinical trial participants are essential to the advancement of medical knowledge and, eventually, the improvement of future patient standards of care.

Researchers painstakingly create a protocol detailing the study's goals, eligibility requirements, treatment strategies, and evaluation techniques before a clinical trial even gets underway. This protocol guarantees that the study is carried out by strict ethical guidelines and scientific standards. It also acts as a guide for the trial. Furthermore, there are usually many stages of clinical trials, each with a distinct function in the drug development process.

Phase I studies evaluate a novel treatment's dose and safety in a limited number of individuals. Phase II studies build on this by assessing the efficacy of the medication and investigating its safety profile in a larger sample size. Phase III studies are the last step in determining if the new therapy is better or non-inferior to current standard therapies in terms of safety and effectiveness.

Patients with pancreatic cancer who may have run out of conventional therapy choices or who want access to potentially ground-breaking medicines may find it proactive to enroll in a research study. But it's crucial to comprehend the advantages and disadvantages of taking part. When thinking about joining a clinical study, patients should carefully explore all of their choices with their medical team and balance any possible benefits against any potential dangers or uncertainties.

Current Research And Breakthroughs

Research on pancreatic cancer has advanced significantly in recent years, offering hope for discoveries and innovative therapeutic strategies. Researchers from all around the globe are working nonstop to understand the nuances of this illness and create more potent diagnostic and therapeutic approaches.

The creation of tailored treatments is one area of interest in pancreatic cancer research. Molecular pathways or genetic abnormalities that fuel the development of cancer are the precise targets of targeted treatments, in contrast to standard chemotherapy, which affects both malignant and healthy cells. Targeted treatments have the potential to provide patients with more precise and less hazardous therapeutic alternatives by focusing on these particular targets.

Another fascinating field of study is immunotherapy, which uses the body's immune system to identify and combat cancer cells. Immunotherapy has proven very effective in treating many forms of cancer, but its effectiveness in treating pancreatic cancer has been more restricted. To overcome these obstacles and realize the full potential of immunotherapy in the fight against pancreatic cancer, however, is the goal of current research.

The development of early detection techniques is another essential area of study attention. Early detection of pancreatic cancer greatly increases the prognosis for long-term survival and effective treatment. To increase early detection and optimize patient outcomes, researchers are investigating novel imaging modalities, biomarkers, and screening procedures.

Furthermore, the development of precision medicine has completely changed the way that cancer is treated by customizing treatments for each patient according to their particular genetic profile and tumor features. Researchers may discover certain mutations or molecular markers that fuel the development of cancer via genomic profiling and molecular analysis, opening the door to more specialized and successful treatment approaches.

Collaboration and knowledge-sharing among researchers, physicians, and patient advocates are crucial for expediting progress and converting scientific findings into concrete benefits for patients, even if the field of pancreatic cancer research is always changing.

How To Participate And Access Experimental Treatments

Patients with pancreatic cancer may find hope and new therapeutic alternatives by taking part in clinical trials and obtaining experimental therapies. It might be intimidating to navigate the process, however, so it's critical to comprehend the procedures and how to take advantage of these changes.

Talking about the possibilities with your medical team is the first step towards taking part in a clinical study. Your oncologist may provide insightful information on current trials, qualifying requirements, possible hazards, and advantages. Based on your medical background and desired course of treatment, they may also assist you in determining whether a clinical trial is a good fit for you.

After finding a clinical trial that piques your interest, you'll have to go through screening to find out whether you qualify. This usually entails a detailed assessment of your health history, present condition, and any particular requirements specified in the study protocol. Before participating in the study, you will be required to provide informed consent if you satisfy the eligibility requirements and want to go on.

It may be more difficult to get experimental therapies outside of clinical trials, although it is nevertheless feasible in certain situations. Expanded access or compassionate use programs are one way to get experimental pharmaceuticals outside of clinical trials for patients with significant or life-threatening diseases. These programs may include extra paperwork and logistical concerns, and they often need clearance from regulatory bodies and the medication manufacturer.

Participating in single-patient or small-group studies run by researchers or academic institutions is another way to get access to experimental medicines. Access to cutting-edge treatments or experimental interventions that are not yet accessible via conventional routes may be provided by these studies. But it's crucial to proceed cautiously when seizing such chances and make sure the therapy is given under the proper medical supervision.

In the end, patients, healthcare professionals, and researchers must work together and give serious thought to taking part in clinical trials or receiving experimental therapies. Even though there may be advantages to these solutions, it's important to consider the pros and cons and make judgments that meet your tastes and treatment objectives. Patients may help develop science and perhaps improve their own and future generations' results by actively participating in the study process.

CHAPTER 8

Survivorship And Quality Of Life

Life After Treatment

Living after pancreatic cancer treatment may be a fresh start, but it may also provide a unique set of difficulties. It's important to keep in mind that care continues even after therapy is over. Maintaining regular follow-up meetings with your healthcare team will remain essential for tracking your health and treating any potential issues.

It's normal to feel a variety of feelings as you adjust to life following treatment, such as relief, worry, and uncertainty. While some survivors may experience appreciation for beating the illness, others could experience dread of a relapse or long-term psychological and physical side effects from the therapy.

It's critical that you allow yourself time to work through these emotions and, if necessary, seek out help from family, friends, support groups, or mental health specialists.

Managing Long-Term Side Effects

Treatment for pancreatic cancer aims to eradicate the illness, but there may be long-term consequences for your body and general health. Fatigue, gastrointestinal problems, neuropathy, changes in appetite and weight, and mental difficulties including anxiety or depression are typical side effects.

It's important to discuss any symptoms or concerns you may have honestly with your healthcare provider to appropriately treat any side effects. They may provide advice on how to manage your symptoms, suggest treatments or lifestyle adjustments to enhance your quality of life, and, if needed, send you to other experts.

A healthy lifestyle may be very helpful in controlling long-term adverse effects and enhancing general well-being, in addition to medical therapies. This might include maintaining a healthy diet, getting regular exercise, using stress-reduction exercises like yoga or meditation, and placing a high value on getting enough sleep.

Maintaining Quality Of Life And Well-Being

For those who have survived pancreatic cancer, maintaining a good quality of life and general well-being is paramount. In addition to managing bodily health, this also entails attending to emotional, social, and spiritual requirements.

For survivors, finding meaning and purpose in life might be especially crucial as they work through the difficulties of being a survivor.

A feeling of purpose may be developed and general well-being can be improved by partaking in joyful and fulfilling activities, such as volunteering, spending time with loved ones, or pursuing hobbies and interests.

During the survival journey, establishing a network of friends, family, and other survivors may be quite beneficial. Making connections with others who have gone through comparable situations may provide comprehension, compassion, and useful guidance for overcoming obstacles and acknowledging accomplishments along the route.

You may create a meaningful and rewarding life after cancer treatment by putting self-care first, getting help when you need it, and continuing to take proactive steps to manage your health.

CHAPTER 9

Caregiver's Guide

Roles And Responsibilities Of Caregivers

When someone receives a pancreatic cancer diagnosis, caring for them becomes very important. Caregivers are essential in helping their loved ones through their cancer journey by offering them practical, emotional, and physical care. Depending on the patient's demands and the disease's stage, a caregiver's duties may change.

Caregivers are there, first and foremost, to provide emotional support and company. They give support and encouragement as well as a shoulder to cry on in trying times. Additionally, caregivers are essential in acting as the patient's advocate and making sure they get the greatest treatment from medical experts.

In practical terms, caregivers help with everyday duties that patients may find difficult to do because of their condition. Assistance with personal cleanliness, food preparation, prescription administration, and travel to doctor's appointments are a few examples of this. Caregivers often develop into specialists in handling the administrative aspects of the patient's care, such as making appointments and liaising with medical professionals.

Caregivers may also be in charge of overseeing the financial elements of cancer treatment. This might include figuring out insurance policies, comprehending medical costs, and looking for support services or financial aid.

In addition to their practical duties, caregivers provide the patient's family and friends with support and information. They could arrange for visits, provide updates on the patient's health, and offer advice on how family members might help them.

All things considered, being a caretaker is a complex and hard job that calls for endurance, compassion, and patience. To avoid burnout and make sure they can keep giving their loved ones adequate care, caregivers must be aware of their limitations and seek help when necessary.

Self-Care Tips For Caregivers

To preserve their well-being, caregivers for a loved one with pancreatic cancer must practice self-care since the disease may be emotionally and physically taxing. Although taking care of oneself might make caregivers feel bad, it's important to understand that doing so is not selfish—rather, it's a must for effectively caring for others.

The most crucial self-care strategy for caregivers is to give their physical health priority. This includes maintaining a healthy diet, getting adequate sleep, and working out often. Resilience and energy levels may be raised by maintaining excellent physical health, which makes it easier for caregivers to handle the responsibilities of caring.

Self-care on an emotional level is as vital. Whether it's reading, outdoor activities, or engaging in a hobby, caregivers should schedule time for things that make them happy and relaxed. Caregivers must also communicate their emotions and ask friends, family, or a support group for assistance. Speaking with someone who can relate to their experiences might help people feel less alone and validate them.

Another essential component of caregivers' self-care is setting limits. It's critical to know when they need a break and to ask for assistance when required. It's okay for caregivers to take time for themselves;

in the end, it helps them to refuel so they can keep giving their loved ones high-quality care.

Caregivers may benefit from practicing stress reduction and mindfulness strategies. Deep breathing techniques, yoga, and meditation are a few practices that may help lower stress levels and foster balance and tranquility.

Ultimately, asking for expert assistance when required is a show of strength rather than weakness. Counseling or therapy may help caregivers process their feelings and create coping mechanisms for handling the difficulties of providing care.

Caregivers may protect their health and provide greater support for their loved one with pancreatic cancer by making self-care a priority.

Support Networks For Caregivers

While providing care for a patient with pancreatic cancer may be a taxing and lonely experience, support networks can provide priceless tools and encouragement. These networks may come from a range of places, such as local organizations, internet forums, and official support groups.

Online groups and forums for caregivers of pancreatic cancer patients are among the easiest places for caregivers to get help. Through these platforms, caregivers may interact with others who can relate to their experiences, provide guidance, and provide emotional support. Online forums and services tailored exclusively for caregivers may be found on websites like CancerCare, PanCAN, and the American Cancer Society.

Another important resource for caregivers is formal support groups. Community centers, religious institutions, or healthcare organizations may serve as facilitators for these gatherings. Caregivers may discuss their worries in a secure environment, gain knowledge from others, and get advice from qualified facilitators in support groups.

Support services for caregivers of cancer patients may also be provided by NGOs and local groups. Financial aid, educational programs, counseling, and respite care are a few examples of these services. Caregivers around the country may access tools and support programs from organizations like the Family Caregiver Alliance and the Cancer Support Community.

Caregivers may get help from friends, family, and neighbors in addition to official support networks. Family members may practically help with caregiving duties, provide emotional support, and provide respite care to give caregivers a much-needed break.

Lastly, caregivers may get great assistance and direction from healthcare experts. Nurses, social workers, and other healthcare professionals who specialize in oncology treatment may give emotional support, counseling, and information about community resources and services.

Caregivers may get the help and resources they need to deal with the difficulties of taking care of a loved one who has pancreatic cancer by connecting with these support networks.

CHAPTER 10

Advocacy And Awareness

Spreading Awareness About Pancreatic Cancer

Raising awareness of pancreatic cancer is essential to combating this often misdiagnosed and underfunded illness. Education is the first step in raising awareness. A lack of knowledge about the signs and risk factors of pancreatic cancer might result in a delayed diagnosis and unfavorable treatment. By making people aware of the warning signs and symptoms of pancreatic cancer, we may encourage them to seek treatment as soon as possible, which may increase their chances of survival.

Participating in outreach initiatives and community activities is one efficient technique to raise awareness. Putting up fundraising events, health fairs, and educational seminars may help spread vital knowledge about pancreatic cancer to a larger audience. Furthermore, we can contact those who would not otherwise be reached by conventional methods and magnify our message by using social media platforms and traditional media outlets.

A crucial component of raising awareness is de-stigmatizing the illness. Because smoking and obesity are two bad lifestyle choices that are often linked to pancreatic cancer, people who have been diagnosed with the illness may have misunderstandings and feel guilty. We can fight stigma and promote empathy in our communities by stressing the complexity of pancreatic cancer and the need to show sympathy and support for individuals who are impacted.

Advocating For Research Funding

To improve our knowledge of pancreatic cancer and provide more potent therapies, we must advocate for funding for research. When compared to other cancer types, pancreatic cancer is disproportionately underfunded, while being one of the deadliest tumors. We can speed up research and enhance patient outcomes by arguing for more financing from public institutions, private foundations, and pharmaceutical businesses.

Organizing grassroots campaigns and lobbying initiatives is one efficient method of promoting financing for research. Funding allocations may be greatly impacted by organizing patients, caregivers, medical professionals, and other interested parties to contact legislators and request that they give pancreatic cancer research top priority. Furthermore, educating political officials on the need to fund pancreatic cancer research may be accomplished via correspondence, meetings, and petitions.

Our lobbying efforts may also be strengthened by working together with patient advocacy groups and advocacy organizations. We can use our combined knowledge and resources to push for more funding and changes to local and national policies by forming alliances with groups and people who share our values. In addition, taking part in occasions like World Pancreatic Cancer Day and Pancreatic Cancer Awareness Month may increase public awareness and provide our advocacy efforts a boost.

Getting Involved In Pancreatic Cancer Organizations

Participating in pancreatic cancer groups is a significant approach to help in the battle against this life-threatening illness. These groups are essential in spreading knowledge, helping patients and their families, and sponsoring studies that can help people with pancreatic cancer live longer and better lives.

Offering your time and talents as a volunteer is one way to get engaged. Numerous groups that fight pancreatic cancer depend on volunteers to help with a variety of tasks, including planning events, doing administrative tasks, and providing peer-to-peer counseling. Through volunteering, you can support the organization's overarching goals and directly impact the lives of individuals impacted by pancreatic cancer.

Taking part in fundraising events is another method to assist pancreatic cancer groups. Every dollar generated, whether via the planning of a charity walk, the holding of a bake sale, or involvement in a peer-to-peer fundraising campaign, goes toward funding important programs and projects. To continue lending the group and its goals continuous financial assistance, you can also think about becoming a corporate sponsor or contributor.

Engaging in advocacy activities via pancreatic cancer groups may also help you make a bigger difference and accelerate significant change. You can push for legislative reforms and more financing for pancreatic cancer support services and research by participating in advocacy campaigns, getting in touch with decision-makers, and spreading awareness in your neighborhood. When we band together, we can make a significant impact in the pancreatic cancer battle.

CONCLUSION

Due to its aggressive nature and poor prognosis, pancreatic cancer requires immediate care as well as novel strategies. In summary, despite tremendous progress in our knowledge of this illness, additional work has to be done.

First and foremost, early detection is a crucial objective. When there are no clear symptoms in the early stages, the diagnosis is sometimes made at an advanced, less curable stage. Investing in biomarker and imaging research has the potential to improve patient outcomes by increasing early identification rates.

Second, improvements in therapy are essential. The poor effectiveness of current treatment options, such as radiation therapy, chemotherapy, and surgery, highlights the critical need for innovative therapeutics.

Precision medicine, immunotherapy, and targeted therapeutics provide promise for individualized care that may improve patient quality of life and survival rates.

Improving supportive care interventions is also critical. Pancreatic cancer affects a person's emotional and psychological well-being in addition to their physical health. Early integration of palliative care into the treatment trajectory may improve patient experience overall, address psychological needs, and reduce symptoms.

To further the fight against pancreatic cancer, cooperation between researchers, physicians, pharmaceutical corporations, and advocacy organizations is crucial. To spur innovation and hasten the creation of efficient preventive, diagnosis, and treatment methods, more money and resources must be devoted to pancreatic cancer research.

In conclusion, there is cause for hope even if pancreatic cancer continues to be one of the most difficult cancers to treat. We may work toward a day where pancreatic cancer is not only a disease that can be successfully avoided, discovered early, and treated with better results, but also one that can be detected with persistence, creativity, and teamwork.

THE END